The Number Puzzler

ROY MULLINS

3	?	17	24	31	?
33	37	30	?	27	31
532	482	427	367	302	?
1	1	2	?	7	13
?	16	54	128	?	432
3	10	29	?	127	218

Tarquin Publications

Logical Thinking, Hidden Connections
Looking for Clues, What comes Next?

This book is a collection of number puzzles on which to test your mind and ingenuity. It is also an interesting and thorough introduction to the methods and techniques which can be used to solve them.

All the puzzles are in the form of a sequence of numbers with missing terms. For each of them there is an underlying rule or hidden connection between the numbers and you have to spot what it could be. When you find the rule, it will explain how the sequence was generated and so make it possible to calculate the missing term or terms. Archimedes leapt from his bath and ran down the street shouting 'Eureka' when he had solved a difficult problem presented to him by the King. With these puzzles we hope that you will experience the same feeling of excitement and satisfaction when you suddenly spot what the pattern is and see how to calculate the answer.

The fascination of puzzles like these, as with so many aspects of mathematics, lies in the combination of logical thinking and sudden bursts of inspiration. There is no single way to solve them but by working through the twelve different types given in this book, you will have gained a valuable insight into suitable methods to try. To reinforce that insight, at the end of each set of examples there is a collection of five puzzles of that type for you to solve yourself.

The final chapter offers five collections of fifteen randomly mixed puzzles, so you will then be able to test your skill and ingenuity to the full. Having mastered them you will be well-placed to solve the number puzzles which are so popular in newspapers, magazines and competitions. Possibly even win some of the prizes which are on offer!

© 2008: Roy Mullins
I.S.B.N: 1 899618 47 3
Design: Magdalen Bear
Printing: Progress Press Co Ltd,
        Malta

Tarquin Publications
99 Hatfield Road
St Albans, Herts
AL1 4JL
United Kingdom
www.tarquinbooks.com

Chapter One

This chapter starts by introducing a most powerful and general systematic method for solving number sequence puzzles known as 'The Method of Extended Differences'. It depends on working out and examining the differences between the successive terms of the sequence. With their help it often becomes possible to work out how the sequence must continue and so to arrive at the solution. This approach is generally applicable as a first line of attack. It should not just be used mechanically. As will be seen, recognising patterns in the differences can often provide important clues to finding other more sophisticated rules for generating the puzzle sequences. Even the lack of an apparently regular pattern can itself be a valuable clue. In this chapter we concentrate on those puzzles where the hidden rules include simple addition, subtraction, multiplication and division.

■ **Type 1:** Adding or subtracting a constant

■ **Type 2:** Adding or subtracting a changing number

■ **Type 3:** Multiplication factors

■ **Type 4:** Division and fractions

Type 1: Adding or subtracting a constant

As might be expected, Type 1 is the most basic type of number-sequence puzzle. In it the step from one term of the sequence to the next is simply to add or subtract a constant number. Example 1.1 is the simplest case of all, being just the natural number sequence, the sequence of counting numbers.

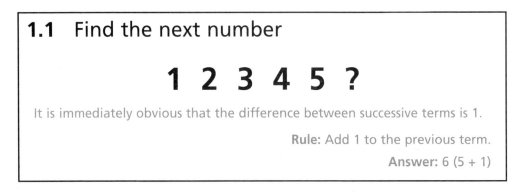

1.1   Find the next number

1 2 3 4 5 ?

It is immediately obvious that the difference between successive terms is 1.

Rule: Add 1 to the previous term.

Answer: 6 (5 + 1)

In this case the sequence is so simple that any number of additional terms can be added almost without having to think at all. However, it does serve as an example of the most important thing to look at with any number sequence puzzle, the differences between successive terms. For reasons that will soon become apparent, these differences are known as the 'first differences'. In this case the first differences are all equal and they are all equal to 1. The other concept is the 'rule' which determines how each term is calculated from the previous one. Essentially, solving a number sequence puzzle is really about discovering what the hidden rule could be.

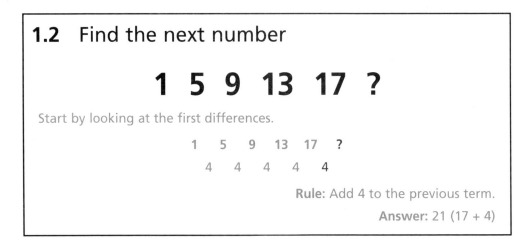

1.2   Find the next number

1 5 9 13 17 ?

Start by looking at the first differences.

1	5	9	13	17	?
	4	4	4	4	4

Rule: Add 4 to the previous term.

Answer: 21 (17 + 4)

In this example the first differences are all equal and are all equal to 4. It can therefore be deduced that the next difference will also be 4 and this is shown in red. It is then clear that the next term is 21.

These next two examples show that sequences can begin with any number and that the first differences can be negative as well as positive.

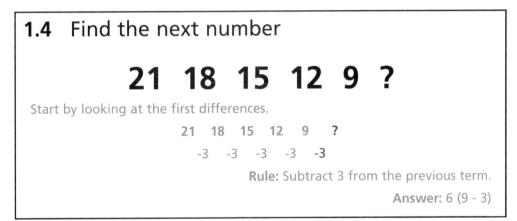

1.3 Find the next number

4 7 10 13 16 ?

Start by looking at the first differences.

4   7   10   13   16   ?

3   3   3   3   3

Rule: Add 3 to the previous term.

Answer: 19 (16 + 3)

1.4 Find the next number

21 18 15 12 9 ?

Start by looking at the first differences.

21   18   15   12   9   ?

-3   -3   -3   -3   -3

Rule: Subtract 3 from the previous term.

Answer: 6 (9 - 3)

Practice Puzzles for Type 1 Sequences

In each case find the rule and the next number.

■ 1a.   **37 56 75 94 ?**

■ 1b.   **84 71 58 45 ?**

■ 1c.   **15 56 97 138 ?**

■ 1d.   **249 227 205 183 ?**

■ 1e.   **472 539 606 673 ?**

Type 2: Adding or subtracting a changing number

The distinctive characteristic of this type of puzzle is that the first differences either increase or decrease in some regular way. Once it can be seen what the next difference must be, the next term of the sequence can be calculated.

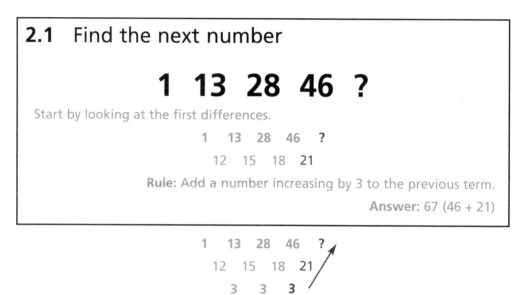

2.1   Find the next number

1 13 28 46 ?

Start by looking at the first differences.

	1	13	28	46	?
	12	15	18	21	

Rule: Add a number increasing by 3 to the previous term.

Answer: 67 (46 + 21)

The idea of 'first differences' can be extended to the calculation of 'second differences' and 'third differences' and so on as long as there are enough terms to calculate them. Here the second differences are all equal to 3. The next first difference is therefore 21 and the answer is 67. This is 'The method of extended differences' in operation and the calculation continues step by step in a kind of diagonal 'ripple effect'.

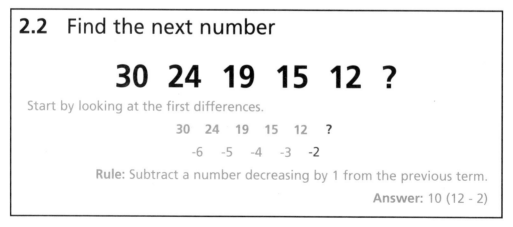

2.2   Find the next number

30 24 19 15 12 ?

Start by looking at the first differences.

	30	24	19	15	12	?
	-6	-5	-4	-3	-2	

Rule: Subtract a number decreasing by 1 from the previous term.

Answer: 10 (12 - 2)

Adding a negative number is the same as subtracting a positive one, so this rule could equally have been given as 'Add a number increasing by 1'.

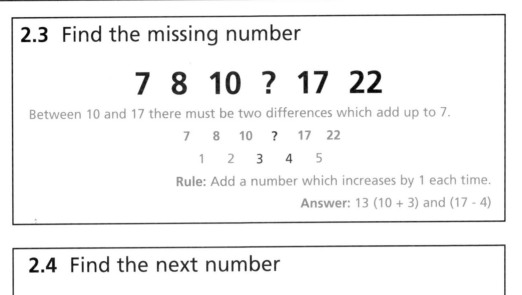

2.3 Find the missing number

7 8 10 ? 17 22

Between 10 and 17 there must be two differences which add up to 7.

| 7 | | 8 | | 10 | | ? | | 17 | | 22 |

| | 1 | | 2 | | 3 | | 4 | | 5 |

Rule: Add a number which increases by 1 each time.

Answer: 13 (10 + 3) and (17 - 4)

2.4 Find the next number

104 120 128 132 134 ?

The first differences show a pattern of halving. It can be continued.

104 120 128 132 134 ?

16   8   4   2   1

Rule: Add a number which is half the previous difference, starting with 16.

Answer: 135 (134 + 1)

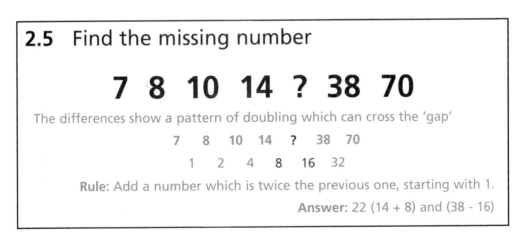

2.5 Find the missing number

7 8 10 14 ? 38 70

The differences show a pattern of doubling which can cross the 'gap'

7   8   10   14   ?   38   70

1   2   4   8   16   32

Rule: Add a number which is twice the previous one, starting with 1.

Answer: 22 (14 + 8) and (38 - 16)

Note that another rule which will produce this same sequence is to double the previous term and subtract 6. Sometimes there may be several different but equivalent ways of formulating the rule and obtaining the answer.

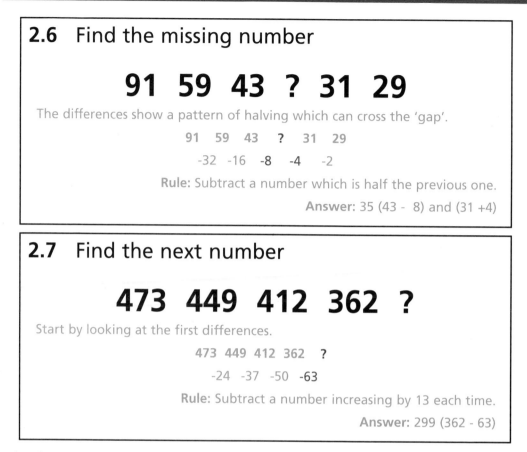

2.6   Find the missing number

91 59 43 ? 31 29

The differences show a pattern of halving which can cross the 'gap'.

91   59   43   ?   31   29

-32 -16 -8   -4   -2

Rule: Subtract a number which is half the previous one.

Answer: 35 (43 - 8) and (31 +4)

2.7   Find the next number

473 449 412 362 ?

Start by looking at the first differences.

473 449 412 362   ?

-24 -37 -50 -63

Rule: Subtract a number increasing by 13 each time.

Answer: 299 (362 - 63)

It has been said that number sequence puzzles such as these belong in the category of 'trial and error' puzzles. However, you will have already seen that this is far from the truth. By continuing to analyse the various types we shall discover which approach is likely to lead to a solution. It could reasonably be called a 'systematic trial' method

Practice Puzzles for Type 2 Sequences

In each case find the rule and the next or omitted number.

■ 2a.   **14 30 50 74 102 ?**   134

■ 2b.   **360 326 295 267 242 ?**   220

■ 2c.   **156 234 310 384 456 ?**   526

■ 2d.   **532 482 427 367 302 ?**   232

■ 2e.   **28 30 33 ? 42 48**   37

Type 3: Multiplication Factors

In the two previous types of sequence each term was obtained from the previous one by adding or subtracting a number. The key difference for this new type is that a factor of multiplication is involved. As we shall see, there may also be a constant to add or subtract. While puzzles of this type can be solved by using differences, there are other simpler and often more interesting rules to be discovered.

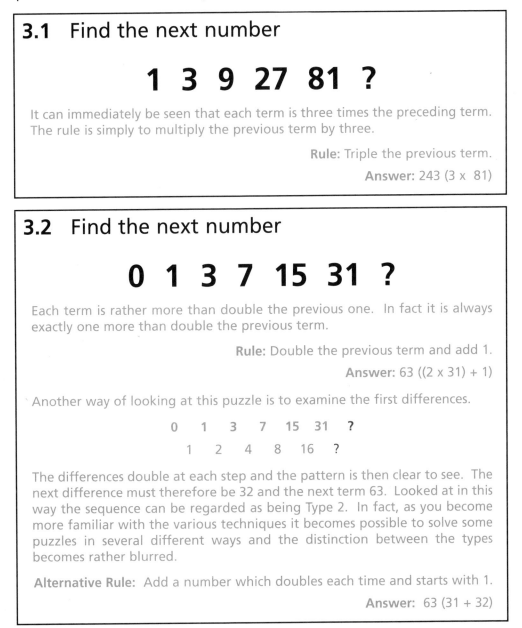

3.1   Find the next number

1 3 9 27 81 ?

It can immediately be seen that each term is three times the preceding term. The rule is simply to multiply the previous term by three.

Rule: Triple the previous term.

Answer: 243 (3 x 81)

3.2   Find the next number

0 1 3 7 15 31 ?

Each term is rather more than double the previous one. In fact it is always exactly one more than double the previous term.

Rule: Double the previous term and add 1.

Answer: 63 ((2 x 31) + 1)

Another way of looking at this puzzle is to examine the first differences.

```
0    1    3    7   15   31    ?
  1    2    4    8   16    ?
```

The differences double at each step and the pattern is then clear to see. The next difference must therefore be 32 and the next term 63. Looked at in this way the sequence can be regarded as being Type 2. In fact, as you become more familiar with the various techniques it becomes possible to solve some puzzles in several different ways and the distinction between the types becomes rather blurred.

Alternative Rule: Add a number which doubles each time and starts with 1.

Answer: 63 (31 + 32)

3.3 Find the next number

5 18 57 174 ?

Start by working out the first differences.

$$5 \quad 18 \quad 57 \quad 174 \quad ?$$
$$13 \quad 39 \quad 117 \quad \mathbf{351}$$

The first thing to notice about the differences is that 39 is three times 13. A little arithmetic shows that 39 x 3 =117, so it seems as if there could be a multiplier of 3. It is also true that 5 x 3 = 15 and 18 x 3 = 54 and 57 x 3 = 171 and so the hidden rule and the next term both become immediately obvious.

Rule: Multiply each term by 3 and add 3.

Answer: 525 ((3 x 174) + 3)

In this example the first differences gave a strong indication of what the multiplier should be. For some more complicated puzzles the number to be added or subtracted may vary according to some rule and then it is harder to identify the multiplier so easily. The next two examples illustrate other ways of tackling the problem.

3.4 Find the next number

8 15 28 53 ?

Start by working out the differences.

$$8 \quad 15 \quad 28 \quad 53 \quad ?$$
$$7 \quad 13 \quad 25 \quad \mathbf{49}$$

There are insufficient terms to identify a pattern in the differences with confidence, so we go back to looking at the original sequence. The last term is almost twice the previous term. Trying this, we find that 15 = (2 x 8) - 1, that 28 = (2 x 15) - 2 and that 53 = (2 x 28) - 3. A pattern becomes apparent and we have a solution. Note that this now gives a second difference sequence of 6, 12, 24,...

Rule: Double each term and subtract a number increasing by 1.

Answer: 102 ((2 x 53) - 4) or (53 + 49)
However, see the important note opposite.

Example 3.4 raises a fundamental point. Although a rule and a solution were found, there could have been another rule and another solution. Taking the second differences not as 6, 12, 24 but as 6, 12, 18, the solution would be 84 and this would be equally valid. To avoid ambiguous situations like this it would normally be the case that a puzzle setter would provide an extra term.

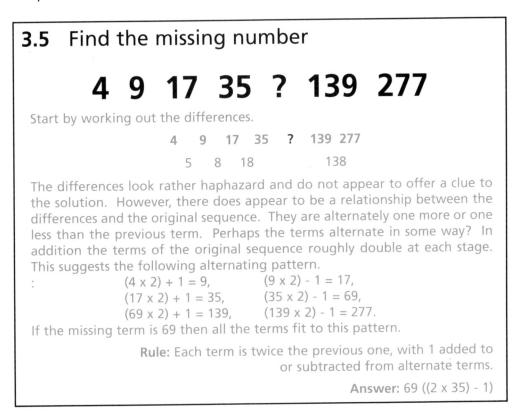

3.5 Find the missing number

4 9 17 35 ? 139 277

Start by working out the differences.

```
         4    9    17   35    ?    139  277
            5    8    18            138
```

The differences look rather haphazard and do not appear to offer a clue to the solution. However, there does appear to be a relationship between the differences and the original sequence. They are alternately one more or one less than the previous term. Perhaps the terms alternate in some way? In addition the terms of the original sequence roughly double at each stage. This suggests the following alternating pattern.

$(4 \times 2) + 1 = 9,$ $(9 \times 2) - 1 = 17,$
$(17 \times 2) + 1 = 35,$ $(35 \times 2) - 1 = 69,$
$(69 \times 2) + 1 = 139,$ $(139 \times 2) - 1 = 277.$

If the missing term is 69 then all the terms fit to this pattern.

Rule: Each term is twice the previous one, with 1 added to or subtracted from alternate terms.

Answer: 69 $((2 \times 35) - 1)$

Practice Puzzles for Type 3 Sequences

In each case find the rule and the next number or numbers.

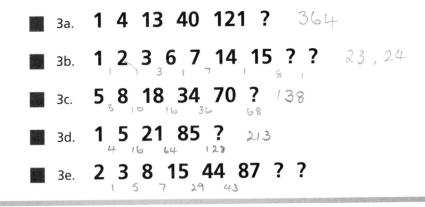

■ 3a. **1 4 13 40 121 ?**   364

■ 3b. **1 2 3 6 7 14 15 ? ?**   23 , 24
       1   3 1 7 1   8 1

■ 3c. **5 8 18 34 70 ?**   138
         3 10 16 36 68

■ 3d. **1 5 21 85 ?**   213
          4 16 64 128

■ 3e. **2 3 8 15 44 87 ? ?**
         1 5 7 29 43

11

Type 4: Division and Fractions

This is a natural extension of the ideas of Type 3 where the multiplication factors are smaller than one and so can be better thought of as division factors. It also includes any multiplication factor which can be conveniently expressed as a fraction.

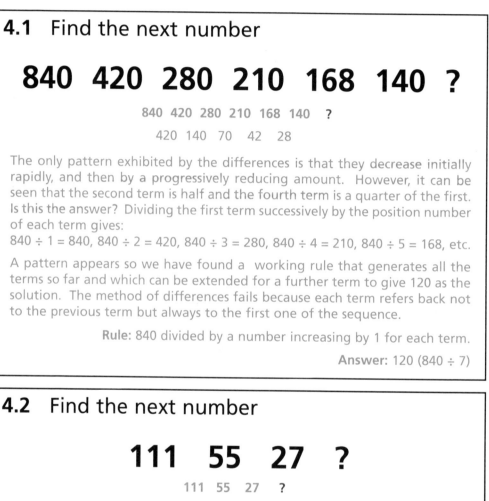

4.1 Find the next number

840 420 280 210 168 140 ?

840 420 280 210 168 140   ?

420 140   70   42   28

The only pattern exhibited by the differences is that they decrease initially rapidly, and then by a progressively reducing amount. However, it can be seen that the second term is half and the fourth term is a quarter of the first. Is this the answer? Dividing the first term successively by the position number of each term gives:

$840 \div 1 = 840$, $840 \div 2 = 420$, $840 \div 3 = 280$, $840 \div 4 = 210$, $840 \div 5 = 168$, etc.

A pattern appears so we have found a working rule that generates all the terms so far and which can be extended for a further term to give 120 as the solution. The method of differences fails because each term refers back not to the previous term but always to the first one of the sequence.

Rule: 840 divided by a number increasing by 1 for each term.

Answer: 120 ($840 \div 7$)

4.2 Find the next number

111   55   27   ?

111   55   27   ?

-56   -28

The differences are negative and 1 bigger than the resulting term.

Looking further at the original sequence, each term is about half the preceding term. In fact, if from each term you subtract 1 and then halve it, it works exactly.

Rule: Subtract 1 from each term and halve it.

Answer: 13 (($27 - 1) \div 2$)

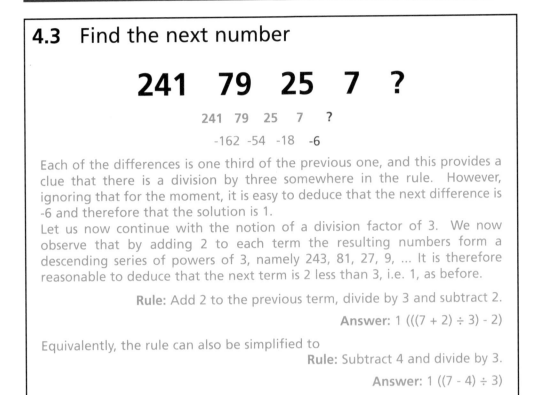

4.3   Find the next number

241 79 25 7 ?

241 79 25 7   ?

-162 -54 -18 -6

Each of the differences is one third of the previous one, and this provides a clue that there is a division by three somewhere in the rule. However, ignoring that for the moment, it is easy to deduce that the next difference is -6 and therefore that the solution is 1.

Let us now continue with the notion of a division factor of 3. We now observe that by adding 2 to each term the resulting numbers form a descending series of powers of 3, namely 243, 81, 27, 9, ... It is therefore reasonable to deduce that the next term is 2 less than 3, i.e. 1, as before.

Rule: Add 2 to the previous term, divide by 3 and subtract 2.

Answer: 1 (((7 + 2) ÷ 3) - 2)

Equivalently, the rule can also be simplified to

Rule: Subtract 4 and divide by 3.

Answer: 1 ((7 - 4) ÷ 3)

4.4   Find the next number

162   108   72   48   ?

162 108 72   48   ?

-54 -36 -24

There is the somewhat awkward pattern that each of the differences is negative and is half of the following term. To demonstrate an alternative line of enquiry, note that these numbers are all composite numbers with several factors. Expressing them in terms of their prime factors, listed on page 56, we have:

162 = 2×3^4, 108 = $2^2 \times 3^3$, 72 = $2^3 \times 3^2$, 48 = $2^4 \times 3^1$ and so   32 = $2^5 \times 3^0$

Note that each term gains a 2 and loses a 3. This gives us a fraction rule.

Rule: Each term is 2/3 of the previous term.

Answer: 32 (48 x 2/3)

4.5   Find the next number

1 3 6 10 15 21 28 ?

1	3	6	10	15	21	28	?
	2	3	4	5	6	7	8

Examination of the first differences provides an immediate solution.

Rule: Add a number increasing by 1 for each term.

Answer: 36 (28 + 8)

Again, in the interests of widening the experience of tackling number sequence puzzles, it can also be regarded in a completely different way. Look at the pairs of successive terms written as fractions.

3/1	6/3	10/6	15/10	21/15	28/21	etc.

Cancelling down in a rather unusual way, we have

3/1	4/2	5/3	6/4	7/5	8/6	etc.

and an alternative rule:

Rule: Multiply the previous term by a fraction, starting with 3/1, whose upper and lower numbers both increase by 1 for each term.

Answer: 36 (28 x 9/7)

This example is a good illustration of how a sequence can have both a very simple rule and a rather complicated and obscure rule and that both can lead to the same answer. While a rule always generates a unique sequence, it does not follow that a given sequence only has one rule. Different rules may generate exactly the same sequence or may generate different sequences with certain terms in common.

Practice Puzzles for Type 4 Sequences

In each case find the rule and the next number.

■ 4a.   **512 384 288 216 ?**

■ 4b.   **3646 1216 406 136 46 ?**

■ 4c.   **629 129 29 9 ?**

■ 4d.   **1024 512 128 16 ?**

■ 4e.   **300,000 90,000 27,000 8,100 ?**

Chapter Two

In Chapter 1 the four types which were dealt with all had one important factor in common. Each new term of the sequence was calculated directly from the preceding term. The rules were such that once the initial term was given, the complete sequence was uniquely determined. This chapter takes the complexity a step further and deals with types of sequence where more than one of the earlier terms are required in order to calculate the next term. There are two distinct types to consider.

■ **Type 5:** Alternating sequences

■ **Type 6:** Depending on more than one previous term

Type 5: Alternating Sequences

This type of puzzle can at first sight be confusing because there seems to be no clear pattern in the numbers. However, once again, differences can help to provide an indication of what to look for. Another clue is that two extra numbers are required.

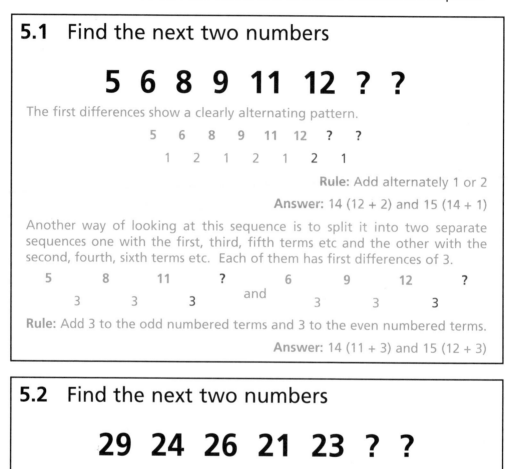

5.1 Find the next two numbers

5 6 8 9 11 12 ? ?

The first differences show a clearly alternating pattern.

5	6	8	9	11	12	?	?
1	2	1	2	1	2	1	

Rule: Add alternately 1 or 2

Answer: 14 (12 + 2) and 15 (14 + 1)

Another way of looking at this sequence is to split it into two separate sequences one with the first, third, fifth terms etc and the other with the second, fourth, sixth terms etc. Each of them has first differences of 3.

5		8		11		?			6		9		12		?
	3		3		3			and		3		3		3	

Rule: Add 3 to the odd numbered terms and 3 to the even numbered terms.

Answer: 14 (11 + 3) and 15 (12 + 3)

5.2 Find the next two numbers

29 24 26 21 23 ? ?

The terms seem neither to be increasing nor decreasing in any clear way. However, the differences clearly show that it is an alternating sequence.

29	24	26	21	23	?	?
-5	2	-5	2	-5	2	

Rule: Alternately subtract 5 and add 2.

Answer: 18 (23 - 5) and 20 (18 + 2)

Alternative Rule: There are two interleaved sequences with differences of -3.

Answer: 18 (21 - 3) and 20 (23 - 3)

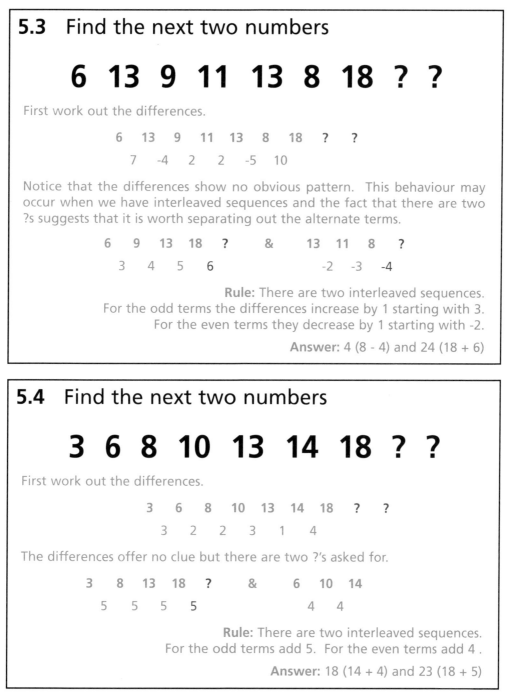

5.3   Find the next two numbers

6 13 9 11 13 8 18 ? ?

First work out the differences.

6   13   9   11   13   8   18   ?   ?
7   -4   2   2   -5   10

Notice that the differences show no obvious pattern. This behaviour may occur when we have interleaved sequences and the fact that there are two ?s suggests that it is worth separating out the alternate terms.

6   9   13   18   ?   &   13   11   8   ?
3   4   5   6   -2   -3   -4

Rule: There are two interleaved sequences.
For the odd terms the differences increase by 1 starting with 3.
For the even terms they decrease by 1 starting with -2.

Answer: 4 (8 - 4) and 24 (18 + 6)

5.4   Find the next two numbers

3 6 8 10 13 14 18 ? ?

First work out the differences.

3   6   8   10   13   14   18   ?   ?
3   2   2   3   1   4

The differences offer no clue but there are two ?'s asked for.

3   8   13   18   ?   &   6   10   14
5   5   5   5   4   4

Rule: There are two interleaved sequences.
For the odd terms add 5. For the even terms add 4 .

Answer: 18 (14 + 4) and 23 (18 + 5)

When presenting the solutions to interleaved sequences it is important to place them in the right order. This means checking to see whether the two required terms are in the order 'even-odd' or 'odd-even'.

Mystery sequences can also be constructed with three or more interleaved elements and then show three or more question marks! Here is a very simple illustrative example.

5.5   Find the next three numbers

2 5 7 4 8 11 6 11 15 ? ? ?

First work out the differences.

	2	5	7	4	8	11	6	11	15	?	?	?
		3	2	-3	4	3	-5	5	4			

These differences offer no clues but as there are three ?'s asked for, let us try testing for three interleaved sequences.

2	4	6	?	&	5	8	11	?	&	7	11	15	?
	2	2	2			3	3	3			4	4	4

Rule: There are three interleaved sequences..
For the first add 2, for the second add 3
and for the third add 4.

Answer: 8 (6 + 2) and 14 (11+ 3) and 19 (15 +4)

Of course with so few terms, it would not require a great deal of ingenuity to construct other plausible rules which would generate the same or different answers.

Practice Puzzles for Type 5 Sequences

In each case find the rule and the next two numbers.

■ 5a.   **32 34 37 39 42 44 ? ?**

■ 5b.   **52 46 47 41 42 36 ? ?**

■ 5c.   **63 86 74 92 85 98 ? ?**

■ 5d.   **8 9 10 14 13 19 17 ? ?**

■ 5e.   **15 8 14 10 12 13 9 17 ? ?**

Type 6: Depending on more than one previous term

Up to now we have been dealing with sequences which have simply progressed from one term to the next according to the given rule. This new type looks at rules which use more than one previous term.

6.1   Find the next number

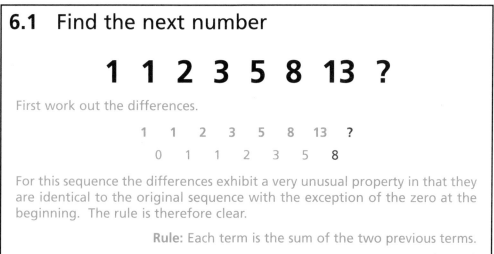

1 1 2 3 5 8 13 ?

First work out the differences.

1	1	2	3	5	8	13	?
0	1	1	2	3	5	8	

For this sequence the differences exhibit a very unusual property in that they are identical to the original sequence with the exception of the zero at the beginning. The rule is therefore clear.

Rule: Each term is the sum of the two previous terms.

Answer: 21 (13 + 8)

6.2   Find the next number

1 2 2 3 4 10 37 ?

1	2	2	3	4	10	37	?
	1	0	1	1	6	27	

The first differences do not immediately indicate a solution. However, we observe that the last two terms are much larger than the earlier ones. So perhaps multiplication rather than addition is involved. One observation is that 4 x 10 = 40 which is close to 37. Also that the preceding term is a 3 and 37 is 3 less than 40. Applying the same rule to the previous term, gives (3 x 4) - 2 = 10 and 4 = (2 x 3) - 2. It works! We cannot go back further because we need to have three terms to start the process.

Rule: Multiply a term by the previous term and subtract the term before that.

Answer: 366 ((10 x 37) - 4)

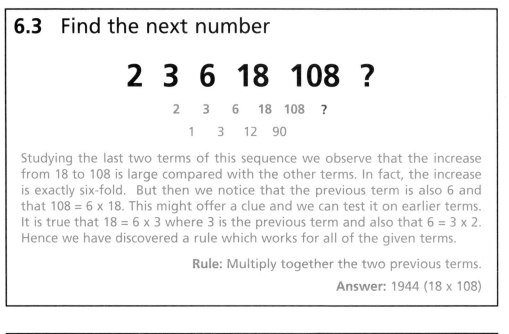

6.3 Find the next number

2 3 6 18 108 ?

2	3	6	18	108	?
1	3	12	90		

Studying the last two terms of this sequence we observe that the increase from 18 to 108 is large compared with the other terms. In fact, the increase is exactly six-fold. But then we notice that the previous term is also 6 and that 108 = 6 x 18. This might offer a clue and we can test it on earlier terms. It is true that 18 = 6 x 3 where 3 is the previous term and also that 6 = 3 x 2. Hence we have discovered a rule which works for all of the given terms.

Rule: Multiply together the two previous terms.

Answer: 1944 (18 x 108)

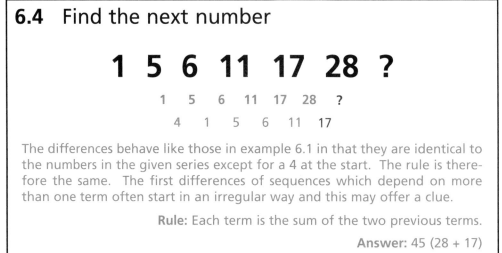

6.4 Find the next number

1 5 6 11 17 28 ?

1	5	6	11	17	28	?
4	1	5	6	11	17	

The differences behave like those in example 6.1 in that they are identical to the numbers in the given series except for a 4 at the start. The rule is therefore the same. The first differences of sequences which depend on more than one term often start in an irregular way and this may offer a clue.

Rule: Each term is the sum of the two previous terms.

Answer: 45 (28 + 17)

Examples 6.1 and 6.4 are examples of an important family of sequences called Lucas Numbers. The first is also called the 'Fibonacci' series which was first described by Leonardo of Pisa in his book *Liber abaci* published in 1202 A.D. He had constructed a rather artificial puzzle about rabbit breeding but it was soon discovered that these 'Fibonacci' numbers also occurred naturally in plant genetics and in other areas of mathematics.

6.5 Find the next number

2 4 4 8 16 64 ?

The differences in this example do not reveal a pattern but it is noticeable that all the numbers are even and are powers of two. A study of successive terms suggests that multiplying them together gives a product which is twice as much as the following one.

Rule: Halve the product of the previous two terms.

Answer: 512 ((16 x 64) ÷ 2)

6.6 Find the next number

52 10 42 32 10 22 ?

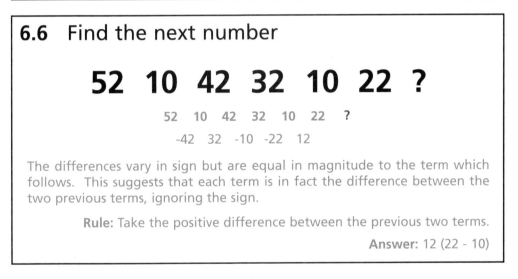

52   10   42   32   10   22   ?

-42   32   -10   -22   12

The differences vary in sign but are equal in magnitude to the term which follows. This suggests that each term is in fact the difference between the two previous terms, ignoring the sign.

Rule: Take the positive difference between the previous two terms.

Answer: 12 (22 - 10)

Practice Puzzles for Type 6 Sequences

In each case find the rule and the next number.

■ 6a. **125 279 404 683 1087 ?**

■ 6b. **2 3 4 10 37 366 ?**

■ 6c. **3 1 4 8 13 25 46 ?**

■ 6d. **1 2 2 4 8 32 ?**

■ 6e. **1 1 9 10 11 20 30 ?**

On this page there are eight sequence puzzles of the first six types and also eight rules which may or may not generate them. The puzzle is to match the rules with the sequences. Is it always possible to match the sequences with the rules? The answers are given on page 52.

The Rules

Rule 1: Subtract 7 from the previous term.

Rule 2: Add a number increasing by 3 to the previous term.

Rule 3: Add 5 to the last but one term.

Rule 4: Add 1 to the previous term.

Rule 5: Divide the previous term by increasing even numbers.

Rule 6: Alternately add 1 to or double the previous term.

Rule 7: Alternately add 2 or 3 to the previous term.

Rule 8: Add the two previous terms together and subtract a number that is increasing by 1.

The Sequences

A. **43 46 52 61 73 88**

B. **3 16 29 42 55 68**

C. **55 48 41 34 27 20**

D. **4 6 9 11 14 16 19**

E. **1920 960 240 40 5**

F. **3 4 8 9 18 19 38 39**

G. **31 32 33 34 35 36**

H. **1 10 19 28 37 46 55**

Chapter Three

For the first two of these three types of puzzles, the crucial point is to recognise that there is a relationship with the sequences of powers of the natural numbers. Fortunately the behaviour of the second and third differences offer a simple method of recognising the presence of powers and offers a clue to any constants or multiplying factors that there are. The last type consists of a miscellaneous group of puzzle sequences which behave in a more diverse but still regular pattern.

■ **Type 7:** Squares and second differences

■ **Type 8:** Cubes and third differences

■ **Type 9:** Other regular sequences

Type 7: Squares and Second Differences

There is an easy way to spot rules which include square numbers. That is to make use of the second differences, the differences between the first differences. It is a fundamental property of sequences involving squares that the second differences are all equal.

7.1 Find the next number

1 4 9 16 25 36 ?

It is probable that you would immediately recognise this sequence as being the squares of the first six natural numbers. See page 55.

Rule: The next square number.

Answer: 49 (7^2)

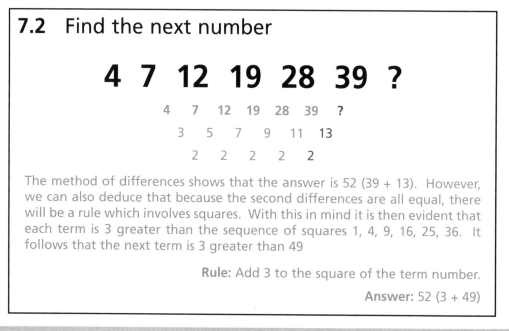

1	4	9	16	25	36	?
	3	5	7	9	11	13
		2	2	2	2	2

The method of differences would have allowed us to calculate the next term without even realising that squares were involved. However, the next puzzle shows how both methods can be used and how they produce the same result.

7.2 Find the next number

4 7 12 19 28 39 ?

4	7	12	19	28	39	?
	3	5	7	9	11	13
		2	2	2	2	2

The method of differences shows that the answer is 52 (39 + 13). However, we can also deduce that because the second differences are all equal, there will be a rule which involves squares. With this in mind it is then evident that each term is 3 greater than the sequence of squares 1, 4, 9, 16, 25, 36. It follows that the next term is 3 greater than 49

Rule: Add 3 to the square of the term number.

Answer: 52 (3 + 49)

The terms of this next example are decreasing, so the relationship with squares is not so immediately obvious.

7.3   Find the next number

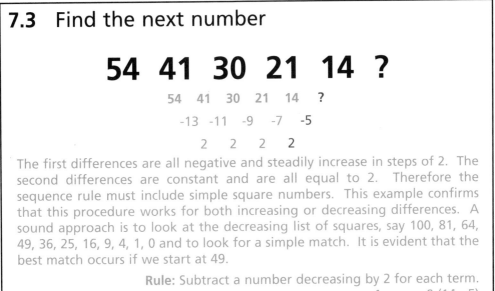

The first differences are all negative and steadily increase in steps of 2. The second differences are constant and are all equal to 2. Therefore the sequence rule must include simple square numbers. This example confirms that this procedure works for both increasing or decreasing differences. A sound approach is to look at the decreasing list of squares, say 100, 81, 64, 49, 36, 25, 16, 9, 4, 1, 0 and to look for a simple match. It is evident that the best match occurs if we start at 49.

Rule: Subtract a number decreasing by 2 for each term.
Answer: 9 (14 - 5)

Rule (Using squares): Add 5 to the next lower square number.
Answer: 9 ($5 + 2^2$)

7.4   Find the next number

142 180 222 268 318 ?

142 180 222 268 318   ?

38   42   46   50   **54**

4   4   4   4

The second differences are all equal and are all equal to 4. This tells us both that a square is involved and also that there is a multiplication factor of 2. For a simple square, the second differences are all 2 and these differences are twice as big. Dividing the numbers of the sequence by 2 removes this multiplying factor and gives a sequence of 71, 90, 111, 134, 159, ? and then we look for a match with the table of squares on page 55. It is then apparent that the 'half' sequence is 10 less than 81, 100, 121, 144, 169.

Rule : Double the next square number starting with 81 and subtract 20.
Answer: 372 ((14^2 x 2) - 20)

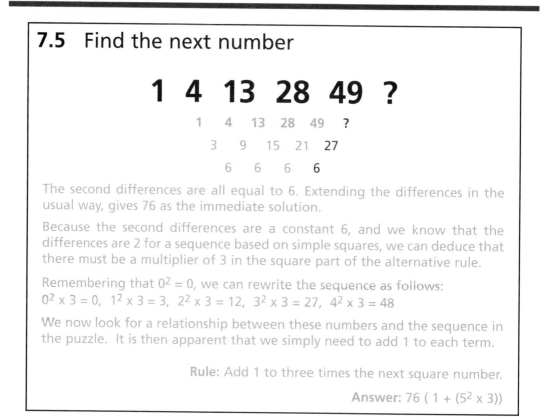

7.5   Find the next number

1 4 13 28 49 ?

1	4	13	28	49	?
3	9	15	21	27	
6	6	6	6		

The second differences are all equal to 6. Extending the differences in the usual way, gives 76 as the immediate solution.

Because the second differences are a constant 6, and we know that the differences are 2 for a sequence based on simple squares, we can deduce that there must be a multiplier of 3 in the square part of the alternative rule.

Remembering that $0^2 = 0$, we can rewrite the sequence as follows:
$0^2 \times 3 = 0$, $1^2 \times 3 = 3$, $2^2 \times 3 = 12$, $3^2 \times 3 = 27$, $4^2 \times 3 = 48$

We now look for a relationship between these numbers and the sequence in the puzzle. It is then apparent that we simply need to add 1 to each term.

Rule: Add 1 to three times the next square number.

Answer: 76 ($1 + (5^2 \times 3)$)

These examples can all be solved by using differences directly and so it almost seems unnecessary to introduce the ideas of squares at all. However, it certainly adds to the interest and to a general sense of wonder about numbers to use differences as a clue to the presence of squares in the underlying rule. These two approaches are further illustrated by the next example.

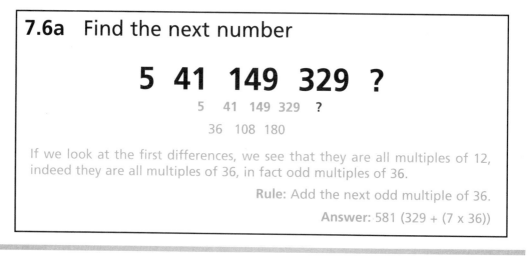

7.6a   Find the next number

5 41 149 329 ?

5	41	149	329	?
	36	108	180	

If we look at the first differences, we see that they are all multiples of 12, indeed they are all multiples of 36, in fact odd multiples of 36.

Rule: Add the next odd multiple of 36.

Answer: 581 (329 + (7 x 36))

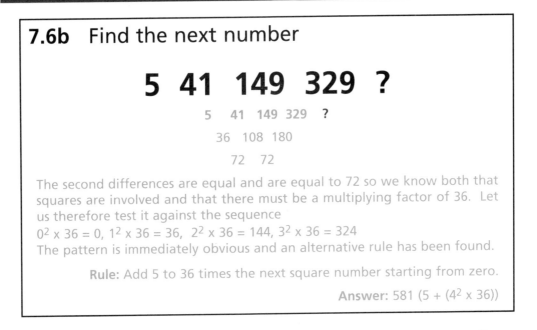

7.6b Find the next number

5 41 149 329 ?

5   41   149 329   ?

36   108 180

72   72

The second differences are equal and are equal to 72 so we know both that squares are involved and that there must be a multiplying factor of 36. Let us therefore test it against the sequence

$0^2 \times 36 = 0$, $1^2 \times 36 = 36$, $2^2 \times 36 = 144$, $3^2 \times 36 = 324$

The pattern is immediately obvious and an alternative rule has been found.

Rule: Add 5 to 36 times the next square number starting from zero.

Answer: 581 ($5 + (4^2 \times 36)$)

The relationship to squares is fortunately easy to remember. If the second differences are equal, then integers to the power of 2 are involved as is a multiplying factor of half its value. There is a similar relationship that if the third differences are equal then integers to the power of 3 are involved. We shall see in the next section how to calculate the multiplication factor. This idea can be carried onwards in principle to fourth, fifth and higher differences. However, there are usually insufficient terms for the method to be of value in solving puzzles like these.

Practice Puzzles for Type 7 Sequences

In each case find the rule and the next number.

■ 7a. **8 11 16 23 32 43 ?**

■ 7b. **46 33 22 13 6 ?**

■ 7c. **1 7 17 31 49 ?**

■ 7d. **5 17 37 65 101 ?**

■ 7e. **0 2 6 12 20 30 42 ?**

Type 8: Cubes and Third Differences

A cube is simply the third power of a number. For example, 7 cubed is written as 7^3 meaning $7 \times 7 \times 7 = 343$. As with squares there are several different ways of solving sequence puzzles with cubes in them. However, for most people the cubes of numbers are less easily recognised than their squares. A list is given on page 55.

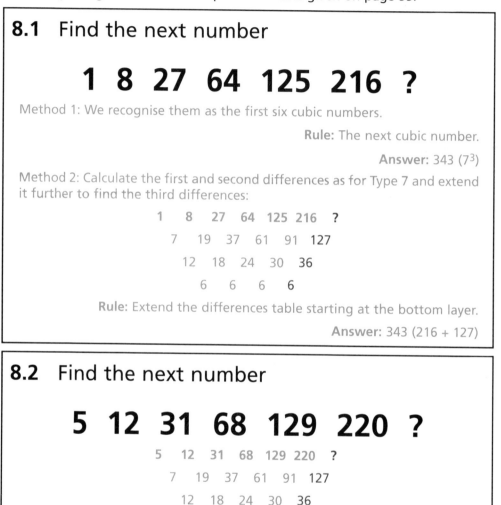

8.1 Find the next number

1 8 27 64 125 216 ?

Method 1: We recognise them as the first six cubic numbers.

Rule: The next cubic number.

Answer: 343 (7^3)

Method 2: Calculate the first and second differences as for Type 7 and extend it further to find the third differences:

	1		8		27		64		125		216		?
		7		19		37		61		91		127		
			12		18		24		30		36			
				6		6		6		6				

Rule: Extend the differences table starting at the bottom layer.

Answer: 343 (216 + 127)

8.2 Find the next number

5 12 31 68 129 220 ?

	5		12		31		68		129		220		?
		7		19		37		61		91		127		
			12		18		24		30		36			
				6		6		6		6				

That the third differences are all equal and are all equal to 6 tells us that this sequence involves simple cubes in some way. In fact, all the differences are the same as before. All we need to do is to add 4 to each cube.

Rule: Add 4 to successive cubes starting with 1.

Answer: 347 (7^3 + 4)

8.3   Find the next number

2 12 36 80 150 252 ?

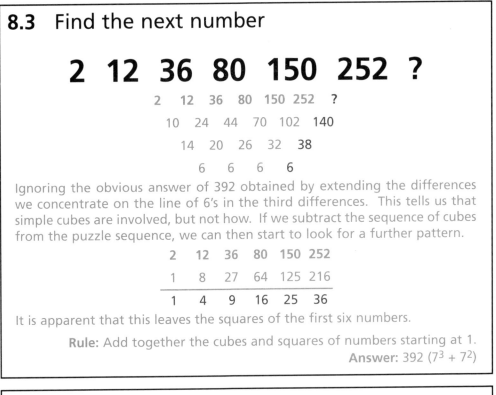

Ignoring the obvious answer of 392 obtained by extending the differences we concentrate on the line of 6's in the third differences. This tells us that simple cubes are involved, but not how. If we subtract the sequence of cubes from the puzzle sequence, we can then start to look for a further pattern.

2	12	36	80	150	252
1	8	27	64	125	216
1	4	9	16	25	36

It is apparent that this leaves the squares of the first six numbers.

Rule: Add together the cubes and squares of numbers starting at 1.

Answer: 392 ($7^3 + 7^2$)

8.4   Find the next number

8 64 216 512 1000 ?

8   64 216 512 1000   ?

56 152 296 488

96 144 192

48   48

The third differences are equal and are equal to 48, indicating that cubes are involved in the sequence and that there is also a multiplying factor of 8. This is because 48 is 8 times the 6 associated with simple cubes. Let us therefore consider the sequence $1^3 \times 8 = 8$, $2^3 \times 8 = 64$, $3^3 \times 8 = 216$, and so on.

Rule: Multiply the cubes of numbers by 8 starting at 1

Answer: 1728 ($6^3 \times 8$)

Alternatively, one can consult the list of cubes on page 55

Rule: The next even cube number

Answer: 1728 (12^3)

This technique can easily be extended to higher powers as the constants of the final row of differences also follow a certain pattern. We have already seen that the natural numbers have equal first differences of 1, the sequence of squares has equal second differences of 2, the sequence of cubes has equal third differences of 6 and it is easy to test that the sequence of fourth powers has equal fourth differences of 24. Notice that these numbers also form a sequence. It is: 1, 2 (1 x 2), 6 (1 x 2 x 3), 24 (1 x 2 x 3 x 4) for quartics, and so on. Such numbers are known as 'factorials' and can be written as 1!, 2!, 3!, 4!, etc. Knowledge of this sequence means that it is a simple matter to identify the multiplication factor of the highest power term in such puzzles.

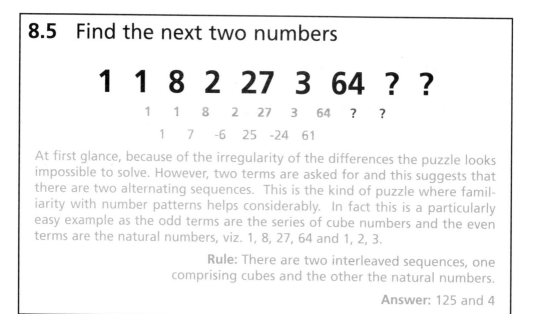

8.5 Find the next two numbers

$$1 \quad 1 \quad 8 \quad 2 \quad 27 \quad 3 \quad 64 \quad ? \quad ?$$

1   1   8   2   27   3   64   ?   ?

1   7   -6   25   -24   61

At first glance, because of the irregularity of the differences the puzzle looks impossible to solve. However, two terms are asked for and this suggests that there are two alternating sequences. This is the kind of puzzle where familiarity with number patterns helps considerably. In fact this is a particularly easy example as the odd terms are the series of cube numbers and the even terms are the natural numbers, viz. 1, 8, 27, 64 and 1, 2, 3.

Rule: There are two interleaved sequences, one comprising cubes and the other the natural numbers.

Answer: 125 and 4

Practice Puzzles for Type 8 Sequences

In each case find the rule and the next number.

■ 8a. **5 12 31 68 129 ?**

■ 8b. **337 210 119 58 21 ?**

■ 8c. **1 27 125 343 ?**

■ 8d. **3 22 59 120 211 ?**

■ 8e. **226 135 74 37 18 ?**

Type 9: Other Regular Sequences

There are obviously many other possibilities for more complex and diverse rules which none the less produce regular sequences. Where the method of differences does not help, we look for relationships between consecutive terms or for relationships with squares, cubes and other powers.

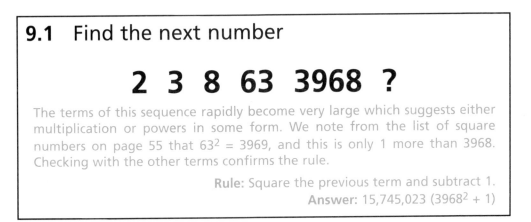

It is interesting to note that, with a sequence of this type, a small change to the rule can produce a very large change in the behaviour of the consecutive terms. As a good illustration of this point, let us modify the rule to: 'Square the previous term and subtract 2'. With an initial term of 2, the sequence becomes 2, 2, 2, 2, 2, ... for ever!

The next two examples show another way in which multiplication might be introduced.

9.2 Find the next number

3 2 2 3 8 35 204 ?

The differences do not show an immediately recognisable pattern, except that successive later terms are increasing rapidly. However, trying various combinations of consecutive numbers reveals that 204 is 6 less than 6 x 35 and that 35 is 5 less than 5 x 8. Checking with the other terms, they behave similarly, and so we have a possible pattern: Multiply the previous term by a number which increases by 1 for each term and then subtract that same number from the product.

This can be simplified slightly, since the effect is the same if we subtract 1 before multiplying. So we have now found a rule which works for all the terms.

Rule: Subtract 1 then multiply by a number increasing by 1 for each term.

Answer: 1421 ((204 - 1) x 7)

9.3 Find the next number

3 6 18 72 360 ?

The differences are no help but the rapidity of increase seems to suggest multiplication. We notice that each term is a multiple of the previous one. For instance 3 x 2 = 6, 6 x 3 = 18 and 18 x 4 = 72. The pattern is then clear.

Rule: Multiply by a number increasing by 1 each time.

Answer: 2160 (360 x 6)

9.4 Find the next number

60 90 108 ?   **Clue:** Degrees

Three terms are insufficient for the differences method and so it must be a trick of some kind. The clue makes one think of angles and so do the numbers 60 and 90. An equilateral triangle has interior angles of 60°, a square 90° and a regular pentagon 108°. The next polygon is a hexagon.

Rule: The interior angles, in degrees, of the next regular polygon.

Answer: 120 (Hexagon)

There is also an arithmetical rule because 60 = 180 x 1/3, 90 = 180 x 2/4 and 108 = 180 x 3/5. The next term is therefore 180 x 4/6 = 120.

Rule: 180 multiplied by a fraction whose top and bottom terms both increase by 1 each time

Answer: 120 (180 x 4/6)

Practice Puzzles for Type 9 Sequences

In each case find the rule and the next number.

■ 9a. **2 10 40 120 240 ?**

■ 9b. **3 24 144 576 ?**

■ 9c. **1 2 5 26 677 ?**

■ 9d. **2310 2 3 5 7 ?**

■ 9e. **2 3 6 11 20 35 ?**

Chapter Four

This chapter covers three types of sequence puzzles where the behaviour is irregular and includes those which are probably the most fun to do. They are ones that remain when all the more systematic and mathematical approaches have failed. In the 1960's Edward de Bono described a process called 'Lateral Thinking' about which he has written a number of interesting books. It is a creative form of thinking from which emerged the concept of an imaginative freewheeling opportunist approach to problem solving. Such thinking often makes use of experience gained in other fields and disciplines. By working through the ingenious examples given in this chapter you will be in a much stronger position when you come to tackle such puzzles on your own. The final type involves prime numbers which in themselves offer no regular pattern to identify. The differences may offer a clue to the presence of prime numbers but the most powerful tool in tackling such puzzles is to become familiar with them. A list of prime numbers below 1000 is given on page 56.

◼ **Type 10:** Alphabet-based sequences

◼ **Type 11:** Oddities and tricks

◼ **Type 12:** Prime numbers

Type 10: Alphabet-based sequences

This type of puzzle appears at first sight to be numerical but is really alphabetical. The given numbers are in fact codes for the letters of the alphabet. One strong indication that this might be the case is that all the numbers are lower than 26.

While many different alphabetical codes are possible, the simplest and the most commonly used code in such puzzles is where the number gives its position in the alphabet. Such numbers can be obtained by counting but it is easier to work from a list.

A = 1, B = 2, C = 3, D = 4, E = 5, F = 6, G = 7, H = 8, I = 9,
J = 10, K = 11, L = 12, M = 13, N = 14, O = 15, P = 16, Q = 17, R = 18,
S = 19, T = 20, U = 21, V = 22, W = 23, X = 24, Y = 25, Z = 26.

This list is also to be found on page 56.

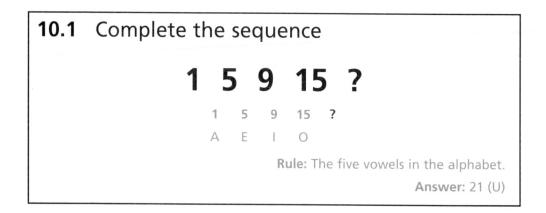

10.1 Complete the sequence

1 5 9 15 ?

1	5	9	15	?
A	E	I	O	

Rule: The five vowels in the alphabet.

Answer: 21 (U)

Once the number sequence is decoded into an alphabetic sequence, it is still a question of spotting a special relationship between the letters. This can sometimes be very difficult and may require some inspired lateral thinking. To help, a clue is often given and this in itself serves as a clue that it is this type of puzzle!

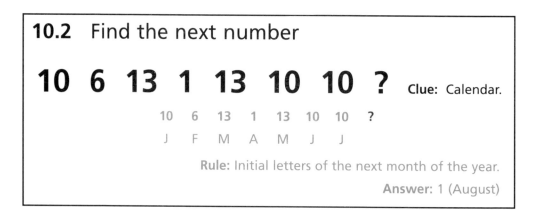

10.2 Find the next number

10 6 13 1 13 10 10 ?   **Clue:** Calendar.

10	6	13	1	13	10	10	?
J	F	M	A	M	J	J	

Rule: Initial letters of the next month of the year.

Answer: 1 (August)

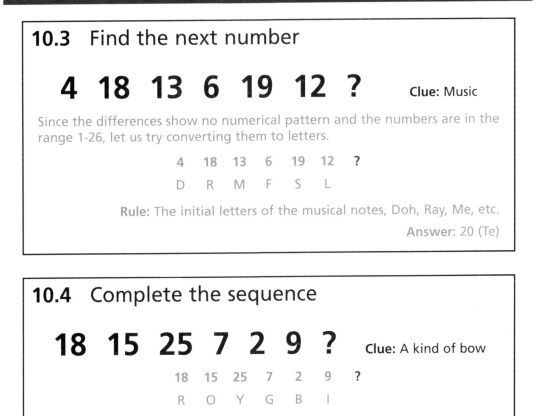

10.3 Find the next number

4 18 13 6 19 12 ?   **Clue:** Music

Since the differences show no numerical pattern and the numbers are in the range 1-26, let us try converting them to letters.

4	18	13	6	19	12	?
D	R	M	F	S	L	

Rule: The initial letters of the musical notes, Doh, Ray, Me, etc.

Answer: 20 (Te)

10.4 Complete the sequence

18 15 25 7 2 9 ?   **Clue:** A kind of bow

18	15	25	7	2	9	?
R	O	Y	G	B	I	

These are the first letters of the colours of the rainBOW.
(Red, Orange, Yellow, Green, Blue, Indigo)

Rule: The initial letter of the last colour of the rainbow.
Answer: 22 (V for violet)

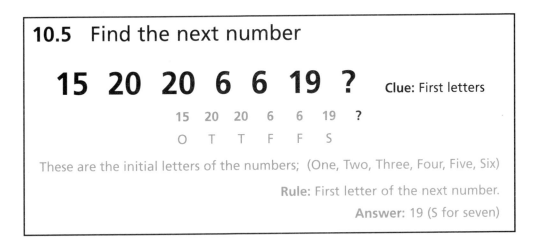

10.5 Find the next number

15 20 20 6 6 19 ?   **Clue:** First letters

15	20	20	6	6	19	?
O	T	T	F	F	S	

These are the initial letters of the numbers; (One, Two, Three, Four, Five, Six)

Rule: First letter of the next number.
Answer: 19 (S for seven)

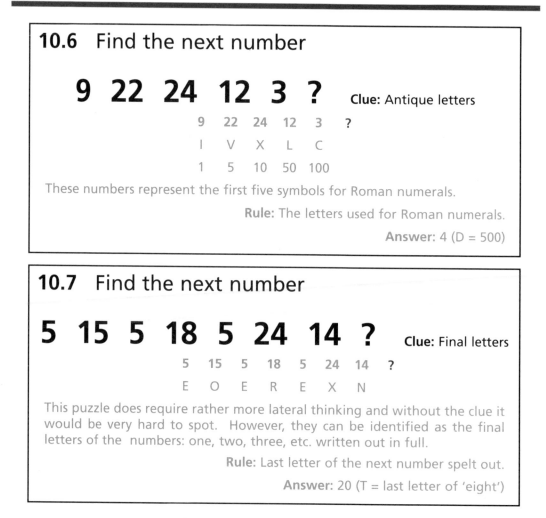

10.6   Find the next number

9 22 24 12 3 ?   **Clue:** Antique letters

9	22	24	12	3	?
I	V	X	L	C	
1	5	10	50	100	

These numbers represent the first five symbols for Roman numerals.

Rule: The letters used for Roman numerals.

Answer: 4 (D = 500)

10.7   Find the next number

5 15 5 18 5 24 14 ?   **Clue:** Final letters

5	15	5	18	5	24	14	?
E	O	E	R	E	X	N	

This puzzle does require rather more lateral thinking and without the clue it would be very hard to spot. However, they can be identified as the final letters of the numbers: one, two, three, etc. written out in full.

Rule: Last letter of the next number spelt out.

Answer: 20 (T = last letter of 'eight')

Practice Puzzles for Type 10 Sequences

In each case find the rule and the next number.

■ 10a. **6 19 20 6 6 19 19 ?**   **Clue:** Ordinal

■ 10b. **18 25 7 2 2 16 ?**   **Clue:** Baize

■ 10c. **17 23 5 18 20 ?**   **Clue:** Keys

■ 10d. **6 18 9 4 1 ?**   **Clue:** Weekday

■ 10e. **2 20 3 20 4 6 5 6 ? ?**   **Clue:** Natural beginning

Type 11: Oddities and tricks

We now introduce some number puzzles which do not fall into any previous category. They are based on ideas such as the numbers of characters in words, the physical shapes of numbers or letters, running familiar sequences of numbers together, inserting spaces in different places and so on in a multitude of ingenious ways devised by clever and imaginative puzzle inventors. Such puzzles offer a similar satisfaction to the pleasure of a good crossword clue. Not only is the answer obvious once it has been thought of but one can also offer a mental nod of appreciation to its constructor. Often such puzzles include a clue because without one, their difficulty might be too great. They have been grouped to show the kinds of approaches which might be required.

11.1   Find the next number

1 8 11 69 88 96 101 ?

Clue: Topsy-turvy

This sequence looks highly irregular and the method of differences therefore offers no clue. Because the number range of the sequence exceeds 26, it is unlikely to be alphabetical. It is true that the first three terms are small while the last four terms are very much larger but this does not seem to help either. However we do have a clue. Topsy-turvy means upside down. It is then obvious that all the given terms would look the same upside down! On further thought, this property applies to no other numbers below 101.

Rule: The next number that looks the same upside-down.

Answer: 111

11.2   Find the next number

1 8 9 13 15 20 21 ?

Clue: Mirror images

Though increasing, they do so irregularly and they all lie between 1 and 26. suggesting that this time it might be an alphabetical code. If so, the letters would be: A H I M O T U. With the help of the clue, we recognise these as letters of the alphabet which look the same in a mirror.

Rule: The next letter which would look the same in a mirror.

Answer: 22 (V)

Because of their mirror-image property, these letters are often used on opticians' eye-test charts.

A simple alphabetic code is not the only way in which a sequence of numbers can be produced from a sequence of letters or words. It is also possible to count the number of letters that a known sequence of words contains. In a rather circular kind of activity, one can count the number of letters in the words for a sequence of numbers!. The following table will be useful for the two examples which follow.

Natural Numbers Letter Count

Three letters	1 2 6 10
Four letters	4 5 9
Five letters	3 7 8 40 50 60
Six letters	11 12 20 30 80 90
Seven letters	15 16 70
Eight letters	13 14 18 19 41 42 46 51 52 56 61 62 66

11.3   Find the next number

3 3 5 4 4 3 5 ?

Clue: Numerically natural

'Numerically natural' suggests the natural numbers: one, two, three, four, etc. and so we look for some kind of connection with them. Since the puzzle sequence only uses numbers in the range 3 to 5 we can be sure that it is not an alphabetical code. In the light of the suggestion above we therefore consider the number of letters in the words for the natural numbers. This immediately offers a satisfactory solution.

Rule: The number of letters in the next natural number.
Answer: 5 (Number of letters in 'eight')

11.4   Find the next number

11 12 20 30 80 ?

Clue: Pick up sticks

The examination of differences and powers offers no suggestion but there is a rather strange clue. 'Pick up sticks' suggests perhaps the children's rhyme which includes '...five, six, pick up sticks'. Perhaps it had something to do with six? A count of the letters in eleven, twelve, twenty etc. shows this to be the case. We therefore look for the next number with six letters. The table above shows that it is in fact the only other such number less than 100.

Rule: The next number with six letters.
Answer: 90 (Ninety)

11.5   Find the next number

1 4 3 11 15 ?   Clue: Spelling

Taking note of the clue and spelling out the numbers given, we see that they are one, four, three, eleven, fifteen and the number of letters is 3, 4, 5, 6, 7. A check will show that each is the lowest number with that number of letters.

Rule: The lowest number with an increasing number of letters.

Answer: 13 (8 letters)

The next two examples illustrate how easily the eye can be confused and connections missed simply by changing the spacing between numbers of the sequence.

11.6   Find the next number

14 91 62 53 64 ?

14   91   62   53   64   ?

77   -29   -9   11

The differences show no regularity and with so few terms it seems impossible to solve arithmetically . We therefore look for another sort of pattern but this time have not been given a clue to help. All the terms have two digits but they do not steadily increase or decrease. Inspiration is required and this comes when we notice that 1, 4, 9 are all squares. This proves to be the decisive idea. The next pairs of digits are 16, 25, 36 and a pattern is found.

Rule: Each term consists of two digits made from successive squares.

Answer: 96 (4<u>9</u> & <u>6</u>4)

11.7   Find the next number

1827 6412 5216 ?

All are four digit numbers and there seems no chance of making use of the differences between them. However 1, 8, 27 are all cubes, as are 64 & 125.

Rule: The next cubic numbers, with digits shown in groups of four.

Answer: 3435 (<u>343</u> & <u>5</u>12)

A similar approach to prime numbers could produce sequences like 235 711 131 719 etc. or 2357 1113 1719 etc. The preponderance of 1's might offer a clue in this case!

The next example is rather special in that it has three different rules all of which lead to the same answer. The first two rules are equivalent and are really both of type 5. The third solution does indeed belong to this chapter of oddities and tricks and that is why it is included here.

11.8 Find the next number

71 42 12 83 54 24 95 ?

Look at the differences

71	42	12	83	54	24	95	?

-29 -30 71 -29 -30 71

Solution 1

These differences show a repeated pattern which it is easy to continue.

Rule: Sequentially subtract 29, subtract 30 and add 71.

Answer: 66 (95 - 29)

Solution 2

Arrange the differences at three different levels

71   42   12   83   54   24   95   ?

Spotting that 83 is 12 greater than 71 and 24 is 12 greater than 12 provides a second solution. This sequence can be regarded as three interleaved sequences each of the simplest type.

Rule: Each term is 12 greater that the term three terms earlier.

Answer: 66 (54 + 12)

Solution 3

Having just seen examples 11.6 and 11.7 a completely different rule is suggested.

Rule: Multiples of 7 split into groups of two digits.

Answer: 66 (56 & 63)

Some ingenuity is required to construct a puzzle such as this because in general, different rules mean that the sequence would continue in a different way and thus produce a different answer. Had three more terms been asked for, then the answers would have been (a) 66, 36, 107, (b) 66, 36, 107 and (c) 66, 37, 07. This rather confirms that (a) and (b) are two ways of treating the same fundamental rule whereas (c) is quite different. You might care to investigate further.

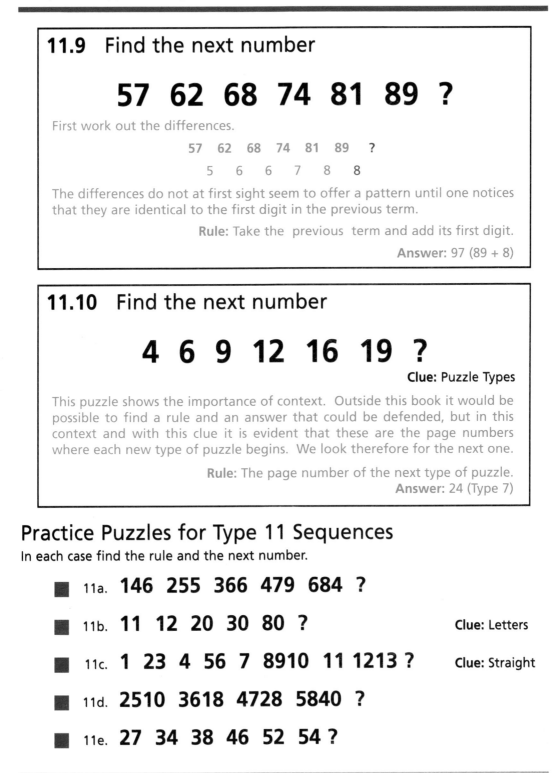

11.9   Find the next number

57 62 68 74 81 89 ?

First work out the differences.

57	62	68	74	81	89	?
5	6	6	7	8	8	

The differences do not at first sight seem to offer a pattern until one notices that they are identical to the first digit in the previous term.

Rule: Take the previous term and add its first digit.

Answer: 97 (89 + 8)

11.10   Find the next number

4 6 9 12 16 19 ?

Clue: Puzzle Types

This puzzle shows the importance of context. Outside this book it would be possible to find a rule and an answer that could be defended, but in this context and with this clue it is evident that these are the page numbers where each new type of puzzle begins. We look therefore for the next one.

Rule: The page number of the next type of puzzle.

Answer: 24 (Type 7)

Practice Puzzles for Type 11 Sequences

In each case find the rule and the next number.

■ 11a. **146 255 366 479 684 ?**

■ 11b. **11 12 20 30 80 ?**   **Clue:** Letters

■ 11c. **1 23 4 56 7 8910 11 1213 ?**   **Clue:** Straight

■ 11d. **2510 3618 4728 5840 ?**

■ 11e. **27 34 38 46 52 54 ?**

Type 12: Prime Numbers

A prime number is one which can be divided only by itself and 1. All are odd with the exception of 2. Familiarity with which numbers are prime often helps to solve such puzzles and a list is given on page 56. Another useful indicator to their presence is that the differences between successive primes are all even. A pattern of differences of 2, 4 or 6 in an irregular sequence suggests a closer look at primes.

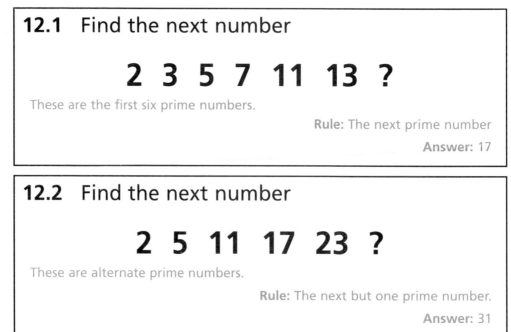

12.1 Find the next number

2 3 5 7 11 13 ?

These are the first six prime numbers.

Rule: The next prime number

Answer: 17

12.2 Find the next number

2 5 11 17 23 ?

These are alternate prime numbers.

Rule: The next but one prime number.

Answer: 31

Some of these practice puzzles are oddities and tricks which involve prime numbers.

Practice Puzzles for Type 12 Sequences

In each case find the rule and the next number.

■ 12a. **22 33 55 77 1111 1313 ?**

■ 12b. **235 711 131 719 232 931 ?**

■ 12c. **4 9 25 49 121 169 289 ?**

■ 12d. **20 30 50 70 110 130 ?**

■ 12e. **29 31 37 41 43 47 ?**

On this page there are eight sequence puzzles of the types 7 to 12 and also eight rules which may or may not generate them. The puzzle is to match the rules with the sequences. Is it always possible to match the sequences with the rules? The answers are given on page 53.

The Rules

Rule 1: Square the previous term and add 2.

Rule 2: Subtract a number increasing by 2 from the previous term.

Rule 3: Alternately, the counting numbers
and then a number increasing by 2.

Rule 4: The position in the alphabet of the letters in a famous composer's name.

Rule 5: 100 minus square numbers, in increasing order.

Rule 6: Square numbers, in increasing order, plus 4.

Rule 7: Alternately, in increasing order,
the square of a prime number and then its cube.

Rule 8: Alternately, in increasing order, the natural numbers and then double the previous term.

The Sequences

A. **5   8   13   20   29   40**

B. **99   96   91   84   75   64**

C. **13   15   26   1   18   20**

D. **3   4   12   48   576**

E. **1   3   11   123**

F. **1   2   2   4   3   6   4   8   5**

G. **216   125   64   27   8   1**

H. **4   8   9   27   25   125   49**

Dealing with the twelve types of puzzle one at a time is an excellent and essential way to acquire the skill and techniques needed to become a master solver of number puzzles like the ones in this book When faced with a new and unknown puzzle this check list will be helpful in deciding how to start to look for a solution.

- Note how many unknown terms are required. If more than one, look for two or even three interleaved sequences.

- Check any wording carefully. If it says 'complete the sequence' then it is probably an alphabetic puzzle or perhaps a numerical trick.

- Work out the first differences and if necessary the second or third. If you can see a simple pattern, then the answer follows immediately.

- If the differences are alternately positive and negative, look for alternating sequences.

- Try multiplying the terms by 2, 3 or 4, or by simple fractions. Are these close to subsequent terms in a regular way?

- Are regular changes slow or rapid? Pay special attention to the largest terms and look for patterns in their factors.

- Note whether the sequence increases, decreases, oscillates systematically or behaves completely irregularly.

- Try adding, subtracting or multiplying two or more previous terms. Can it produce an answer close to the next term?

- Look for numbers on or close to familiar values of prime numbers and to squares, cubes, or other powers of the natural numbers.

- An irregular sequence with numbers in the range 1 to 26 may be based on alphabet substitution.

- An irregular sequence with numbers in the range 3 to 7 may be based on word-lengths.

- Check whether numbers have been run together or have had their digits reversed.

- Look at the physical shape of numbers and their alphabetic equivalents.

Chapter Five

The 75 puzzles in this chapter are for you to solve and they are arranged in the form of five random sets of fifteen. You have to discover for yourself what type each of them is and what might be a suitable method to use. The guidelines in the form of a check list opposite may be helpful in deciding what approaches to try.

■ **Final Challenge:** Mixed Fifteen A

■ **Final Challenge:** Mixed Fifteen B

■ **Final Challenge:** Mixed Fifteen C

■ **Final Challenge:** Mixed Fifteen D

■ **Final Challenge:** Mixed Fifteen E

Mixed Fifteen

A miscellaneous collection of puzzles for you to try.
In each case find the missing term or terms.

■ A1. **3 10 17 24 ?**

■ A2. **2 3 7 14 24 ?**

■ A3. **33 37 30 34 27 31 ? ?**

■ A4. **1 9 19 47 113 ?**

■ A5. **54 45 37 ? 24 19 15**

■ A6. **3 6 11 18 27 38 ?**

■ A7. **1 3 4 7 11 18 ? ?**

■ A8. **3 5 7 11 13 17 19 ?**

■ A9. **3 13 113 1113 ?**

■ A10. **14 19 5 ?**   **Clue:** Compass

■ A11. **1 1 2 4 7 13 24 ?**

■ A12. **16 29 42 55 ?**

■ A13. **1 2 4 7 ? 16 22**

■ A14. **2 6 14 30 62 ?**

■ A15. **2 4 7 9 12 14 ? ?**

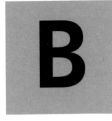

Mixed Fifteen

A miscellaneous collection of puzzles for you to try.
In each case find the missing term or terms.

■ B1. **67 79 74 86 81 93 ? ?**

■ B2. **59 48 37 26 ?**

■ B3. **12 21 33 48 66 ?**

■ B4. **2 8 18 32 50 72 ?**

■ B5. **2 5 7 12 19 31 ?**

■ B6. **16 21 46 171 ?**

■ B7. **223 224 228 ? 245 258 274**

■ B8. **1 4 7 11 14 17 41 ?**

■ B9. **17 13 11 7 5 3 ?**

■ B10. **481 216 202 428 323 ?**

■ B11. **51 44 38 33 29 ?**

■ B12. **13 22 5 13 10 19 21 ?**   **Clue:** Celestial

■ B13. **129 146 163 180 ?**

■ B14. **1 5 7 17 31 ? ?**

■ B15. **2 9 28 65 126 217 ?**

Mixed Fifteen

A miscellaneous collection of puzzles for you to try. In each case find the missing term or terms.

C1. **1 4 6 14 26 ? ?**

C2. **121 132 154 187 ?**

C3. **1 2 3 4 6 7 10 11 15 ? ?**

C4. **5 7 11 31 71 91 32 ?**

C5. **15 20 30 ? 65 90**

C6. **342 215 124 63 26 ?**

C7. **359 322 285 248 ?**

C8. **3 9 19 33 51 73 ?**

C9. **2 12 20 4 ?**   Clue: Meals

C10. **6 8 12 14 18 20 24 30 ?**

C11. **112 77 49 ? 14 7**

C12. **0 3 8 15 24 35 ?**

C13. **219 250 265 296 311 342 ? ?**

C14. **2 3 5 6 8 9 20 ?**

C15. **1 3 2 3 3 5 4 4 ? ?**

Mixed Fifteen

A miscellaneous collection of puzzles for you to try.
In each case find the missing term or terms.

■ D1. **2 4 11 23 40 ?**

■ D2. **14 7 15 8 16 9 ? ?**

■ D3. **3 7 13 27 53 ? ?**

■ D4. **49 41 34 28 ? 19 16**

■ D5. **11 36 61 86 ?**

■ D6. **4 23 13 25 4 3 ?**   Clue: Periods of time

■ D7. **2 6 18 54 162 ?**

■ D8. **2 16 54 128 250 432 ?**

■ D9. **3 9 7 49 11 121 15 ?**

■ D10. **2 10 26 50 82 122 ?**

■ D11. **1 2 5 10 20 50 ?**   Clue: Purse

■ D12. **8 11 16 23 32 43 ?**

■ D13. **22 1 23 24 47 71 ?**

■ D14. **3 10 29 66 127 218 ?**

■ D15. **19 8 4 ?**   Clue: Suits

Mixed Fifteen

A miscellaneous collection of puzzles for you to try. In each case find the missing term or terms.

E1. **253 432 611 790 969 ?**

E2. **1 2 4 9 23 64 ?**

E3. **61 67 71 73 79 83 89 ?**

E4. **8 25 4 18 15 7 5 ?** Clue: Sun

E5. **7 18 35 60 113 208 ?**

E6. **409 401 385 353 ?**

E7. **26 37 50 65 ?**

E8. **32 30 39 27 47 22 56 15 ? ?**

E9. **10 11 100 101 110 ?**

E10. **347 374 437 473 734 ?**

E11. **1 2 6 24 120 ?**

E12. **221 348 517 734 ?**

E13. **7 9 200 216 425 ?**

E14. **7777 1297 217 37 ?**

E15. **29 95 53 ? 61 18 87**

Rules
Answers
& Lists

The solutions and rules which are given here are correct in the sense that they do generate the puzzle sequences in question. However, as we have seen throughout this book, there may be other rules which would also generate the same sequences.

From a mathematician's point of view, many different rules can be constructed which would generate any given finite sequence. Most of them would be very complicated indeed. However, from the point of view of a Number Puzzler, what we are looking for are relatively simple straightforward rules and a unique solution. Often several different rules can be constructed which are logically equivalent and generate the same sequence.

It is an important quality of a good number puzzle that its constructor has made certain that only one relatively simple answer exists. A lot of the satisfaction of finding the solution is lost if there should be other equally valid and simple answers too. It is a feature of this book that most are 'well-constructed' in this sense but with the important exceptions of those like Examples 3.4 and 11.8. These ambiguous puzzles were included in order to make a serious point. That sometimes two different rules are in fact just two different ways of looking at the same sequence and that at other times the two rules generate the given part of the sequence but then diverge afterwards. Where this happens there are perfectly sensible alternative solutions which can be defended and accepted.

■ **Pages 52 to 54:** Solutions and their rules

■ **Pages 55 to 56:** Lists of useful numbers

Bearing in mind the important comments overleaf here are the expected rules and solutions. There may be different rules leading to the same solution.

Practice Puzzles: Type 1 (page 5)
1a. **113** (94 + 19)
    Rule: Add 19 to the previous term.
1b. **32** (45 - 13)
    Rule: Subtract 13 from the previous term.
1c. **179** (138 + 41)
    Rule: Add 41 to the previous term.
1d. **161** (183 - 22)
    Rule: Subtract 22 from the previous term.
1e. **740** (673 + 67)
    Rule: Add 67 to the previous term.

Practice Puzzles:Type 2 (page 8)
2a. **134** (102 + 32)
    Rule: Add a number increasing by 4.
    (16, 20, 24, 28)
2b. **220** (242 - 22)
    Rule: Subtract a number decreasing by 3.
    (34, 31, 28, 25)
2c. **526** (456 + 70)
    Rule: Add a number decreasing by 2.
    (78, 76, 74, 72)
2d. **232** (302 - 70)
    Rule: Subtract a number increasing by 5.
    (50, 55, 60, 65)
2e. **37** (33 + 4) & (42 - 5)
    Rule: Add a number increasing by 1.

Practice Puzzles: Type 3 (page 11)
3a. **364** ((121 x 3) + 1)
    Rule: Multiply by 3 and add 1.
3b. **30** (15 x 2) and **31** (30 + 1)
    Rule: Alternately double or add 1.
3c. **138** ((70 x 2) - 2) & **278** ((138 x 2) + 2)
    Rule: Each term is double the previous term, with 2 added or subtracted alternately.
3d. **341** ((85 x 4) + 1)
    Rule: Multiply by 4 and add 1.
3e. **260** ((87 x 3) - 1) and **519** ((260 x 2) - 1)
    Rule: Alternately, double and subtract 1, then treble and subtract 1.

Practice Puzzles: Type 4 (page 14)
4a. **162** (216 x 3/4)
    Rule: Each term is 3/4 of the previous term.
4b. **16** ((46 + 2) ÷ 3)
    Add 2 and divide by 3 to get the next term.
4c. **5** ((9 - 4) ÷ 5) + 4)
    Rule: Subtract 4 , divide by 5, add 4.
4d. **1** ($16 ÷ 2^4$)
    Rule: Divide each term by an increasing power of 2. (2, 4, 8, 16)
4e. **2430** (8100 x 3/10)
    Rule: Multiply previous term by 3/10.

Practice Puzzles: Type 5 (page 18)
5a. **47** (44 + 3) & **49** (47 + 2)
    Rule: Add 2 or 3 alternately.
5b. **37** (36 + 1) & **31** (37 - 6)
    Rule: Subtract 6 or add 1 alternately.
5c. **96** (85 + 11) & **104** (98 + 6)
    Rule: There are two interleaved sequences, one adding 11 and the other adding 6 to the previous term in their sequence.
5d. **24** (19 + 5) & **22** (17 + 5)
    Rule: There are two interleaved sequences. For one, add 5 to each term. For the other add a number increasing by 1.(2, 3, 4)
5e. **5** (9 - 4) & **22** (17 + 5)
    Rule: There are two interleaved sequences. For one, subtract a number increasing by 1. (1, 2, 3) For the other, add a number increasing by 1. (2, 3, 4)

Practice Puzzles - Type 6 (page 21)
6a. **1770** (683 + 1087)
    Rule: Each term is the sum of the two previous terms.
6b. **13532** ((37 x 366) - 10)
    Rule: Multiply the two previous terms and subtract the term before that.
6c. **84** (13 + 25 + 46)
    Rule: Each is the sum of the three previous terms.
6d. **256** (8 x 32)
    Rule: Each term is the product of the two previous terms.
6e. **41** (30 + 11)
    Rule: Each term is the sum of the previous term and the term two places before that.

Match the Rules Puzzles (page 22)
Rule: 1	Sequence C	
Rule: 2	Sequence A	
Rule: 3	Sequence D	
Rule: 4	Sequence G	No rules are included
Rule: 5	Sequence E	for Sequences B and H
Rule: 6	Sequence F	
Rule: 7	Sequence D	
Rule: 8	Sequence G	

Practice Puzzles: Type 7 (page 27)
7a. **56** ($7^2 + 7$)
    Rule: Add 7 to the next square number.
7b. **1** ($2^2 - 3$)
    Rule: Subtract 3 from each decreasing square number.
7c. **71** ($2 x 6^2$) - 1)
    Rule: Subtract 1 from twice the next square number.
7d. **145** ($12^2 + 1$)
    Rule: Add 1 to the next even square number.
7e. **56** ($8^2 - 8$)
    Rule: The square of the next number, minus the number itself.

Practice Puzzles: Type 8 (page 30)

8a. **220** ($6^3 + 4$)
Rule: Add 4 to the next cube.

8b. **2** ($2^2 - 6$)
Rule: Subtract 6 from the next decreasing cube.

8c. **729** (9^3)
Rule: The next odd cube.

8d. **338** ($7^3 - 5$)
Rule: Subtract 5 from the next cube.

8e. **11** ($1^3 + 10$)
Rule: Add 10 to the next decreasing cube.

Practice Puzzles: Type 9 (page 32)

9a. **240** (240 x 1)
Rule: Multiply the previous term by a decreasing number. (5, 4, 3, 2)

9.b **1152** (576 x 2)
Rule: Multiply the previous term by a decreasing even number. (8, 6, 4)

9c. **458330** ($677^2 + 1$)
Rule: Square the previous term and add 1.

9d. **11**
Rule: 2310 followed by its set of prime factors.

9e. **60** ((20 + 35) + 5)
Rule: Total of the two previous terms, plus a number increasing by 1 for each term.

Practice Puzzles: Type 10 (page 36)

10a. **5** (Eighth)
Rule: The position in the alphabet of the first letter of each of the ordinal numbers. (First, etc.)

10b. **2** (Black)
Rule: The position in the alphabet of the first letter of the colours of snooker balls in value order. (Red, Yellow, Green, Brown, Blue, Pink)

10c. **25** (QWERTY)
Rule: The position in the alphabet of the letters on the top row of typewriters or computer keyboards.

10d. **25** (Y)
Rule: The position in the alphabet of the letters in the word FRIDAY.

10e. **6 & 19** (6, Six)
Rule: The natural numbers followed by the position in the alphabet of their first letters.

Practice Puzzles: Type 11 (page 41)

11a. **891** (9 in 8(-)1)
Rule: The next number in the midst of its square.

11b. **90** (Ninety)
Rule: Those numbers with six letters in order.

11c. **14**
Rule: The counting numbers written in groups, so that those made up of only straight lines are separated and the others are run together.

11d. **6954**
Rule: The first pair of digits increases by 11, the second pair is the product of the first pair.

11e. **58** (54 + 4)
Rule: Take the previous term and add its last digit.

Practice Puzzles: Type 12 (page 42)

12a. **1717**
Rule: The next prime number written twice.

12b. **374** (37 & 4(1))
Rule: Prime numbers with the digits written in groups of three.

12c. **361** (19^2)
Rule: The next prime number squared.

12d. **170** (17 x 10)
Rule: The next prime number multiplied by 10.

12e. **53**
Rule: The next prime number.

Match the Rules Puzzles (page 43)

Rule: 1	Sequence E	
Rule: 2	Sequence B	
Rule: 3	Sequence F	
Rule: 4	Sequence C	No rules are included
Rule: 5	Sequence B	for Sequences D and G
Rule: 6	Sequence A	
Rule: 7	Sequence H	
Rule: 8	Sequence F	

The Five Final Challenges

Bearing in mind the important comments on page 51, here are the expected rules and solutions. There may be different rules leading to the same solution.

Mixed Fifteen A (page 46)

A1. **31** (24 + 7) **Rule:** Add 7 to each term.

A2. **37** (24 + 13) **Rule:** Add a number increasing by 3.

A3. **24** (31 - 7) and **28** (24 + 4) **Rule:** Add 4 and subtract 7 alternately.

A4. **273** ((113 x 2) + 47) **Rule:** Double the previous term and add the second previous term.

A5. **30** (37 - 7) **Rule:** Subtract a number decreasing by 1 from the previous term.

A6. **51** ($7^2 + 2$) **Rule:** Square numbers plus 2.

A7. **29** (11 + 18) and **47** (18 + 29) **Rule:** Each term is the sum of the two previous terms.

A8. **23** **Rule:** The next prime number.

A9. **11113** ($3 + 10^4$) **Rule:** Add 3 to the powers of 10.

A10. **23** (West) **Rule:** The position in the alphabet of the first letters of the four compass points.

A11. **44** (7 + 13 + 24) **Rule:** Add the three previous terms.

A12. **68** (55 + 13) **Rule:** Add 13 to each term.

A13. **11** (7 + 4) **Rule:** Add a number increasing by 1 to each term.

A14. **126** (62 + 64) **Rule:** Add to each term a number which doubles each time.

A15. **17** (14 + 3) and 19 (17 + 2) **Rule:** Add 2 and 3 alternately.

Mixed Fifteen B (page 47)

B1. **88** (93 - 5) and **100** (88 + 12) **Rule:** Add 12 and subtract 5 alternately.

B2. **15** (26 - 11) **Rule:** Subtract 11 from each term.

B3. **87** (66+ 21) **Rule:** Add a number increasing by 3.

B4. **98** (7^2 x 2) **Rule:** Double the next square number.

B5. **50** (19 + 31) **Rule:** Each term is the sum of the two previous terms

B6. **796** (171 + 5^4) **Rule:** Add the next increasing power of 5.

B7. **235** (228 + 7) **Rule:** Add a number increasing by 3 to each term.

B8. **44 Rule:** Numbers formed of straight lines only.

B9. **2 Rule:** A decreasing sequence of primes.

B10.**640** (36 & 40) **Rule:** The four-times table written in groups of three digits.

B11.**26** (29 - 3) **Rule:** Subtract a number decreasing by 1 from each term.

B12.**14** (Neptune) **Rule:** The position in the alphabet of the first letter of the names planets going outwards from the sun.

B13.**197** (180 + 17) **Rule:** Add 17 to each term.

B14.**65** ((31 x 2) + 3) & **127** ((65 x 2) - 3) **Rule:** Double each term then alternately add 3 or subtract 3.

B15.**344** (7^3 + 1) **Rule:** Add 1 to the next cube.

Mixed Fifteen C (page 48)

C1. **54** ((26 x 2) + 2) & **106** ((54 x 2) - 2) **Rule:** Double each term and alternately add 2 or subtract 2.

C2. **231** (187 + 44) **Rule:** Add a number increasing by 11 to each term.

C3. **16** (11 + 5) & **21** (15 + 6) **Rule:** There are two interleaved sequences. For each one add a number increasing by 1 to the previous term in their sequences.

C4. **92 Rule:** The prime numbers with digits reversed.

C5. **45** (30 + 15) **Rule:** Add a number increasing by 5 to each term.

C6. **7** (8 - 1) **Rule:** Decreasing cubic numbers minus 1.

C7. **211** (248 - 37) **Rule:** Subtract 37 from each term.

C8. **99** ((7^2 x 2) + 1) **Rule:** Double the next square number and add 1.

C9. **19** (Supper). **Rule:** The position in the alphabet of the first letter of the names of daily meals.

C10.**32** (31 + 1) **Rule:** Prime numbers plus 1.

C11.**28** (49 - 21) **Rule:** Subtract a number decreasing by 7 from each term.

C12.**48** (7^2 - 1) **Rule:** Square numbers minus 1.

C13.**357** (342 + 15) & **388** (357 + 31) **Rule:** Add 31 and 15 alternately.

C14.**22 Rule:** The numbers which do not include a digit whose shape is made up only of straight lines.

C15.**5** (4 + 1) & **4** (Five) **Rule:** The odd terms are the natural numbers and the even terms are the number of letters in that previous number.

Mixed Fifteen D (page 49)

D1. **62** (40 + 22) **Rule:** Add to the previous term a number increasing by 5.

D2. **17** (9 + 8) & **10** (17 - 7) **Rule:** Subtract 7 and add 8 alternately.

D3. **107** ((53 x 2) + 1) & **213** (107 x 2) - 1) **Rule:** Double each term and alternately add 1 or subtract 1.

D4. **23** (28 - 5) **Rule:** Subtract a number decreasing by 1 from each term.

D5. **111** (86 + 25) **Rule:** Add 25 to each term.

D6. **13** (M for Millennium) **Rule:** The position in the alphabet of the first letter of time periods.

D7. **486** (162 x 3) **Rule:** Triple each number.

D8. **686** (7^3 x 2) **Rule:** Double the next cube.

D9. **225** (15^2) **Rule:** Add 4 to each odd term and square the odd term to give the next even term.

D10.**170** (122 + 48) **Rule:** Add a number increasing by 8 to each term.

D11.**100 Rule:** The value in pence of the next coin.

D12.**56** (49 + 7) **Rule:** Add 7 to the next square number.

D13.**118** (47 + 71) **Rule:** Each term is the sum of the two previous terms.

D14.**345** (7^3 + 2) **Rule:** Add 2 to the next cube.

D15.**3** (C for Clubs) **Rule:** The position in the alphabet of the first letter of suits of playing cards.

Mixed Fifteen E (page 50)

E1. **1148** (969 + 179) **Rule:** Add 179 to each term.

E2. **186** ((64 x 3)-6) **Rule:** Multiply the previous term by 3, then subtract a number starting with 1 and increasing by 1 for each term.

E3. **97 Rule:** The next prime number.

E4. **14** (N) **Rule:** The position in the alphabet of the letters in Hydrogen (the most abundant element in the sun).

E5. **381** (60 + 113 +208) **Rule:** Each term is the sum of the previous three terms.

E6. **289** (353 - 64) **Rule:** Subtract a number starting with 8 and doubling for each term.

E7. **82** (9^2 + 1) **Rule:** Add 1 to the next square.

E8. **66** (56 + 10) & **6** (15 - 9) **Rule:** There are two interleaved sequences. For one, add a number increasing by 1 for the other subtract a number increasing by 2.

E9. **111 Rule:** The binary numbers.

E10. **743 Rule:** Final rearrangement of the digits.

E11. **720** (120 x 6) **Rule:** Factorials.

E12. **1005** (10^3 x+ 5) **Rule:** Add 5 to the next cube.

E13. **841** (200 + 216 + 425) **Rule:** Each term is the sum of the three previous terms.

E14. **7** (((37 -1) ÷ 6) + 1) **Rule:** Subtract 1 from each term then divide by 6 and add 1.

E15. **36** (53 & 61) **Rule:** The last digit of the previous term and the first digit of the next term.

■ Squares: 1 to 100

$1^2 = 1$	$21^2 = 441$	$41^2 = 1681$	$61^2 = 3721$	$81^2 = 6561$
$2^2 = 4$	$22^2 = 484$	$42^2 = 1764$	$62^2 = 3844$	$82^2 = 6724$
$3^2 = 9$	$23^2 = 529$	$43^2 = 1849$	$63^2 = 3969$	$83^2 = 6889$
$4^2 = 16$	$24^2 = 576$	$44^2 = 1936$	$64^2 = 4096$	$84^2 = 7056$
$5^2 = 25$	$25^2 = 625$	$45^2 = 2025$	$65^2 = 4225$	$85^2 = 7225$
$6^2 = 36$	$26^2 = 676$	$46^2 = 2116$	$66^2 = 4356$	$86^2 = 7396$
$7^2 = 49$	$27^2 = 729$	$47^2 = 2209$	$67^2 = 4489$	$87^2 = 7569$
$8^2 = 64$	$28^2 = 784$	$48^2 = 2304$	$68^2 = 4624$	$88^2 = 7744$
$9^2 = 81$	$29^2 = 841$	$49^2 = 2401$	$69^2 = 4761$	$89^2 = 7921$
$10^2 = 100$	$30^2 = 900$	$50^2 = 2500$	$70^2 = 4900$	$90^2 = 8100$
$11^2 = 121$	$31^2 = 961$	$51^2 = 2601$	$71^2 = 5041$	$91^2 = 8281$
$12^2 = 144$	$32^2 = 1024$	$52^2 = 2704$	$72^2 = 5184$	$92^2 = 8464$
$13^2 = 169$	$33^2 = 1089$	$53^2 = 2809$	$73^2 = 5329$	$93^2 = 8649$
$14^2 = 196$	$34^2 = 1156$	$54^2 = 2916$	$74^2 = 5476$	$94^2 = 8836$
$15^2 = 225$	$35^2 = 1225$	$55^2 = 3025$	$75^2 = 5625$	$95^2 = 9025$
$16^2 = 256$	$36^2 = 1296$	$56^2 = 3136$	$76^2 = 5776$	$96^2 = 9216$
$17^2 = 289$	$37^2 = 1369$	$57^2 = 3249$	$77^2 = 5929$	$97^2 = 9409$
$18^2 = 324$	$38^2 = 1444$	$58^2 = 3364$	$78^2 = 6084$	$98^2 = 9604$
$19^2 = 361$	$39^2 = 1521$	$59^2 = 3481$	$79^2 = 6241$	$99^2 = 9801$
$20^2 = 400$	$40^2 = 1600$	$60^2 = 3600$	$80^2 = 6400$	$100^2 = 10000$

■ Cubes: 1 to 25

$1^3 = 1$	$6^3 = 216$	$11^3 = 1331$	$16^3 = 4096$	$21^3 = 9261$
$2^3 = 8$	$7^3 = 343$	$12^3 = 1728$	$17^3 = 4913$	$22^3 = 10648$
$3^3 = 27$	$8^3 = 512$	$13^3 = 2197$	$18^3 = 5832$	$23^3 = 12167$
$4^3 = 64$	$9^3 = 729$	$14^3 = 2744$	$19^3 = 6859$	$24^3 = 13824$
$5^3 = 125$	$10^3 = 1000$	$15^3 = 3375$	$20^3 = 8000$	$25^3 = 15625$

■ Powers: 0 to 5

$2^0 = 1$	$2^1 = 2$	$2^2 = 4$	$2^3 = 8$	$2^4 = 16$	$2^5 = 32$
$3^0 = 1$	$3^1 = 3$	$3^2 = 9$	$3^3 = 27$	$3^4 = 81$	$3^5 = 243$
$4^0 = 1$	$4^1 = 4$	$4^2 = 16$	$4^3 = 64$	$4^4 = 256$	$4^5 = 1024$
$5^0 = 1$	$5^1 = 5$	$5^2 = 25$	$5^3 = 125$	$5^4 = 625$	$5^5 = 3125$
$6^0 = 1$	$6^1 = 6$	$6^2 = 36$	$6^3 = 216$	$6^4 = 1296$	$6^5 = 7776$
$7^0 = 1$	$7^1 = 7$	$7^2 = 49$	$7^3 = 343$	$7^4 = 2401$	$7^5 = 16807$
$8^0 = 1$	$8^1 = 8$	$8^2 = 64$	$8^3 = 512$	$8^4 = 4096$	$8^5 = 32768$
$9^0 = 1$	$9^1 = 9$	$9^2 = 81$	$9^3 = 729$	$9^4 = 6561$	$9^5 = 59049$

■ Powers: 6 to 8

$2^6 = 64$	$2^7 = 128$	$2^8 = 256$
$3^6 = 729$	$3^7 = 2187$	$3^8 = 6561$
$4^6 = 4096$	$4^7 = 16384$	$4^8 = 65536$

■ Prime Numbers: 1 to 1000

2	61	149	239	347	443	563	659	773	887
3	67	151	241	349	449	569	661	787	907
5	71	157	251	353	457	571	673	797	911
7	73	163	257	359	461	577	677	809	919
11	79	167	263	367	463	587	683	811	929
13	83	173	269	373	467	593	691	821	937
17	89	179	271	379	479	599	701	823	941
19	97	181	277	383	487	601	709	827	947
23	101	191	281	389	491	607	719	829	953
29	103	193	283	397	499	613	727	839	967
31	107	197	293	401	503	617	733	853	971
37	109	199	307	409	509	619	739	857	977
41	113	211	311	419	521	631	743	859	983
43	127	223	313	421	523	641	751	863	991
47	131	227	317	431	541	643	757	877	997
53	137	229	331	433	547	647	761	881	
59	139	233	337	439	557	653	769	883	

■ Composite numbers: 4 to 100

$4 = 2^2$	$26 = 2 \times 13$	$46 = 2 \times 23$	$65 = 5 \times 13$	$85 = 5 \times 17$
$6 = 2 \times 3$	$27 = 3^3$	$48 = 2^4 \times 3$	$66 = 2 \times 3 \times 11$	$86 = 2 \times 43$
$8 = 2^3$	$28 = 2^2 \times 7$	$49 = 7^2$	$68 = 2^2 \times 17$	$87 = 3 \times 29$
$9 = 3^2$	$30 = 2 \times 3 \times 5$	$50 = 2 \times 5^2$	$69 = 3 \times 23$	$88 = 2^3 \times 11$
$10 = 2 \times 5$	$32 = 2^5$	$51 = 3 \times 17$	$70 = 2 \times 5 \times 7$	$90 = 2 \times 3^2 \times 5$
$12 = 2^2 \times 3$	$33 = 3 \times 11$	$52 = 2^2 \times 13$	$72 = 2^3 \times 3^2$	$91 = 7 \times 13$
$14 = 2 \times 7$	$34 = 2 \times 17$	$54 = 2 \times 3^3$	$74 = 2 \times 37$	$92 = 2^2 \times 23$
$15 = 3 \times 5$	$35 = 5 \times 7$	$55 = 5 \times 11$	$75 = 3 \times 5^2$	$93 = 3 \times 31$
$16 = 2^4$	$36 = 2^2 \times 3^2$	$56 = 2^3 \times 7$	$76 = 2^2 \times 19$	$94 = 2 \times 47$
$18 = 2 \times 3^2$	$38 = 2 \times 19$	$57 = 3 \times 19$	$77 = 7 \times 11$	$95 = 5 \times 19$
$20 = 2^2 \times 5$	$39 = 3 \times 13$	$58 = 2 \times 29$	$78 = 2 \times 3 \times 13$	$96 = 2^5 \times 3$
$21 = 3 \times 7$	$40 = 2^3 \times 5$	$60 = 2^2 \times 3 \times 5$	$80 = 2^4 \times 5$	$98 = 2 \times 7^2$
$22 = 2 \times 11$	$42 = 2 \times 3 \times 7$	$62 = 2 \times 31$	$81 = 3^4$	$99 = 3^2 \times 11$
$24 = 2^3 \times 3$	$44 = 2^2 \times 11$	$63 = 3^2 \times 7$	$82 = 2 \times 41$	$100 = 2^2 \times 5^2$
$25 = 5^2$	$45 = 3^2 \times 5$	$64 = 2^6$	$84 = 2^2 \times 3 \times 7$	

■ Alphabet conversion

A	B	C	D	E	F	G	H	I	J	K	L	M
1	2	3	4	5	6	7	8	9	10	11	12	13

N	O	P	Q	R	S	T	U	V	W	X	Y	Z
14	15	16	17	18	19	20	21	22	23	24	25	26

If you have enjoyed this book there may be other Tarquin books which would interest you, including 'The Number Detective' by Jon Millington. Tarquin books are available from bookshops, toy shops and gift shops or in case of difficulty, directly by post from the publishers.

See our full range of books on our secure website at **www.tarquinbooks.com**

Alternatively, if you would like our latest printed catalogue please contact us by email:

info@tarquinbooks.com phone: 0870 143 2568 or write to us at Tarquin Publications, 99 Hatfield Road, St Albans, Herts, AL1 4JL, United Kingdom.